Kari Jormakka
with Oliver Schürer and
Dörte Kuhlmann

Design Methods

Kari Jormakka
with Oliver Schürer and
Dörte Kuhlmann

Design Methods

BIRKHÄUSER
BASEL

Contents

Foreword

Design is a heterogeneous process – approaches, strategies and methodologies are often influenced by the designer's own experiences and socio-cultural background, as well as by technical and economic conditions. On the one hand, design draws on individual creative power, while on the other, it is grounded in methodological principles that reflect a variety of basic attitudes and processes.

Building on the results of *Basics Design Ideas* – which deals primarily with the inspiration and initial stimulus for the design process – this book presents design methods that are governed by rules and not based primarily on intuition. The authors' objective is to provide readers with a range of methods and to motivate them to examine familiar concepts and architects in greater detail. The book describes approaches from geometry and the natural environment, from music and mathematics, from unconscious and rationalist sources, and from generative processes. Well-known buildings are used as examples to explain these methods: the plans and sections of these buildings are analyzed to show how the architects have derived specific solutions. Since the didactic concept of this book is largely based on examples from architectural history – thereby diverging from the concept of other *Basics* books – it does without typical stylistic elements of the series in favor of weaving a common thread throughout the essay.

Adapted to the structure of architectural studies, *Basics Design Methods* primarily targets advanced architectural students and graduates who wish to learn more about design methods. The book does not aim to disseminate any one design method, but to provide a series of practical design tools that can be used to solve design assignments based on specific needs.

Bert Bielefeld, Editor

Introduction

Most poets pretend that they compose by ecstatic intuition, and would positively shudder at letting the public take a peep behind the scenes – or so Edgar Allan Poe claims in his 1846 essay on "The Philosophy of Composition." Instead of mystifying the creative act, Poe candidly reveals how he composed his most famous poem, "The Raven." By his detailed exposition he wants to demonstrate that nothing in the darkly Romantic poem is referable to accident or intuition and that the work proceeded, step by step to its completion, with the precision and strict consequence of a mathematical proof.

Poe's method of composition has been the inspiration for many later writers, composers, artists – and architects, as well. But why should anyone follow a specific method to come up with an architectural design? Some architects claim we need such a method because the problems today are too complex to solve by unaided intuition or traditional wisdom. Others expect that a proper method will enable them to make objectively correct decisions. There are also those who recommend rigorous methods in order to prevent architecture from degenerating into a self-indulgent celebration of the architect's own personality – that is, a private language – or a thoughtless reproduction of familiar models. Some avant-garde methods play with the idea that the role of the architect is reduced by letting certain decisions be made according to chance, while others involve future users in the actual design process.

This book examines various methods of designing architecture, with the help of examples, to determine the strengths and weaknesses of each. Many of them were developed in recent decades, while others have been part of the architect's toolbox for centuries. Although many theorists have stated that they were presenting a universal method that was applicable to all buildings throughout the world, there is good reason to argue that no one particular method can be declared the only correct one for every task. It is thus important to choose a method that is most suitable to the challenge of a particular assignment. Familiarity with several methods offers the designer the most flexibility. But a method is not a machine to solve architectural problems automatically: it focuses, but does not curtail, the real work of solving design challenges.

Nature and geometry as authorities

BIOMORPHIC ARCHITECTURE

Originally the question of a design method concerned the generation of form. Central to the modernist program was the claim that the forms of historical architecture no longer corresponded to the spirit of the age: the old styles had degenerated into an immoral, anachronistic masquerade that hampered the creativity of architects, sent reactionary and dishonest messages, and failed to meet the challenges of the new social and technological conditions.

The architects who were determined to be modern, as architect and theorist Claude Bragdon observed in 1915, identified three main sources of a new architectural language: original genius, nature, and geometry. The reliance on genius may be illustrated by Antoni Gaudí's Casa Milà (1907) in Barcelona and August Endell's Atelier Elvira (1897) in Munich.
> Figs. 1 and 2

However, many architects felt that such experiments were too subjective and whimsical to replace the authority of the past. They wanted to ground architecture on a basis more universal than the caprice of an individual designer, more timeless than changing fashions, and more general than local customs. The study of nature provided models that would be understandable and valid in different societies, irrespective of

Fig. 1: Antoni Gaudí, roofscape of Casa Mila, Barcelona

Fig. 2: August Endell, facade of Atelier Elvira, Munich

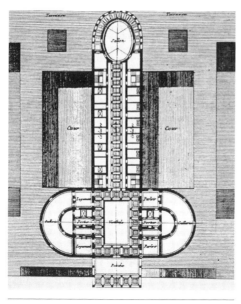

Fig. 3: Claude-Nicolas Ledoux, project for a brothel

historical and political contingencies, while geometry promised access to what may be even more invariable: namely, the principles of order and the laws of thought. Hence, in their attempt to avoid imitating historical precedents, early modern architects often turned to models taken from nature or to the sciences in general to find new shapes for buildings.

Many historical ornaments were derived from the shapes of plants or animals: the classical Corinthian capital features acanthus leaves, and the classical bucranium ornament is in the form of an ox skull. At the end of 18th century, certain architects who argued for *l'architecture parlante,* a "speaking architecture" that referred more or less directly to the objects for which the building was intended, took this idea to extremes: Jean-Jacques Lequeu designed a dairy building in the form of a cow and Claude-Nicolas Ledoux gave the floor plan for a brothel a phallic shape. > Fig. 3

The use of such iconic signs was intended to form a natural language of architecture that would make the function of the buildings understandable across centuries and latitudes, but the more radical designs of "speaking architecture" were never built.

Nonetheless, organicism returned at the end of the 19th century. For example, in 1905 H. P. Berlage designed a chandelier in the form of a

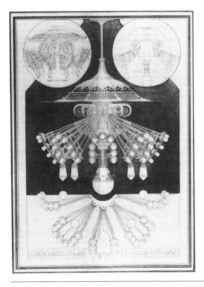

Fig. 4: Hendrik Petrus Berlage, lamp in the shape of a jellyfish

Fig. 5: Hector Guimard, metro entrance in Paris

jellyfish (as illustrated in Ernst Haeckel's *Kunstformen der Natur*). And at about the same time, Hector Guimard imitated the shapes of flowers and insects in his designs for entrances to the Paris metro stations.
> Figs. 4 and 5

The anthroposophist Rudolf Steiner seems to have mixed metaphors in the design of the boiler house (1915) for his Society's commune in Dornach, Switzerland, merging plant leaves with an overall phallic form. Another expressionist architect from that time, Hermann Finsterlin, evoked the shapes of jellyfish, mussels and amoebae in his highly idiosyncratic, unbuilt designs in the early 1920s. > Figs. 6 and 7

Even later, architects occasionally returned to shapes directly suggesting plants or animals. A case in point is the TWA terminal (1956–62) at the JFK International Airport, New York, designed by Eero Saarinen: to signal its function as the entryway to the airplanes, it was shaped like a bird ready to take flight.

Such direct borrowings from the natural world have also been criticized. Instead of imitating the shapes as such, many architects have opted for the imitation of nature in more abstract terms. Already the oldest surviving treatise on architectural theory, the *Ten Books on Architecture* (c. 46–30 BC) by Roman architect Vitruvius, recommends applying

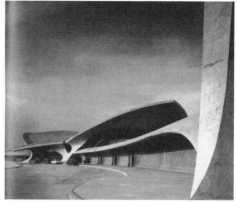

Fig. 6: Rudolf Steiner, boiler house in Dornach

Fig. 7: Eero Saarinen, TWA terminal in New York

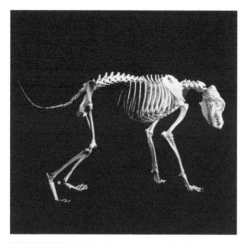

Fig. 8: Dog skeleton

Fig. 9: Santiago Calatrava, structure inspired by a dog skeleton (project)

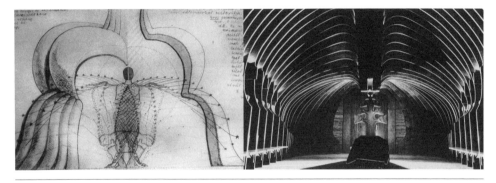

Fig. 10: Imre Makovecz, Farkasrét-Kapelle in Budapest

human proportions in buildings without imitating any of the specific shapes of the human body. Later, architects often studied organisms in order to develop optimum structural shapes. For example, when Santiago Calatrava was commissioned to design an extension to the Cathedral Church of St. John the Divine in New York, he derived inspiration from the skeleton of dog. The final design is a synthesis of two fundamentally different considerations: on the one hand, the evocation of organic shapes; on the other, structural performance. > Fig. 8

The funeral chapel in Farkasrét, Hungary (1975), designed by Imre Makovecz, illustrates a different way of how the shapes of natural organisms can be adjusted to make architectural sense. The highly articulate roof structure was derived from a chronophotograph of Makovecz waving his arms. Here, the photographic technique retains the complex geometry of the body, while providing an image abstract enough to make a reasonable structure. > Figs. 9 and 10

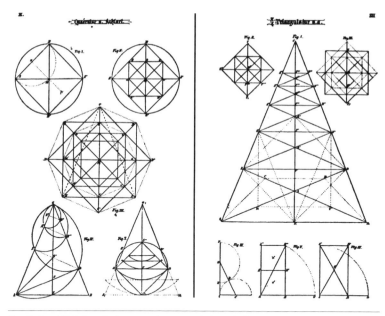

Fig. 11: Hendrik Petrus Berlage, quadrature and triangulation

QUADRATURE AND TRIANGULATION

Another attempt to escape the traps of architectural convention involved scientific models and mathematical procedures. For instance, in his mature work, instead of organic models, Berlage usually worked with proportional systems and geometrical grids to determine his forms with precision. In his writings he discusses two methods from Gothic architecture, known as "quadrature" and "triangulation". > Figs. 11 and 12

In general, quadrature is a mathematical method of determining the area of a plane figure by dividing it into a collection of shapes, the overall area of which is known. In architecture, however, quadrature refers to a specific method of doubling or halving the area of a known square. For example, if we have a square, then a new one, half its size, can be easily drawn by connecting the middle points on each side with lines at an angle of 45°. Triangulation involves a similar method, usually based on an equilateral triangle.

One reason why the early modernists were fascinated by quadrature and triangulation is that these design methods were part of the mythical

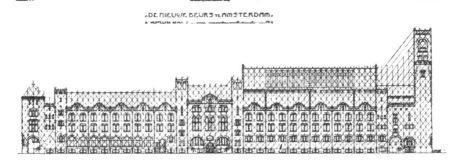

Fig. 12: Hendrik Petrus Berlage, Stock Exchange in Amsterdam

"secret of the masons" in the Middle Ages. Gothic architects used these techniques mainly for pragmatic reasons. Itinerant masons could not use scale drawings, as there was no generally agreed system of measurement – the length of the foot varied from country to country, even from town to town. Instead, they used geometry as a tool for deriving the measurements of the building from a scaleless sketch without a yardstick. Although quadrature and triangulation were largely matters of expediency, they produced highly complex architecture that was coherent and harmonious in its proportions.

Louis Sullivan, one of the pioneers of modern architecture, explained his own geometric methods in his 1924 book *A System of Architectural Ornament*. Starting with a simple square intersected by diagonal and orthogonal axes, he applied quadrature and other geometrical operations to ultimately arrive at delicate floral motifs, which gradually cover up the basic square. Moreover, Sullivan argued that he recognized in these organic forms a feminine principle emerging out of the dominant male principle of geometric order. The transcendentalist idea, according to which life grows out of such opposing powers and that the universe rests on a dualist foundation, formed the conceptual basis for his ornamental designs. > Fig. 13

Later modernists suppressed such symbolic readings but often continued to rely on geometry. Frank Lloyd Wright, who once worked as an assistant to Sullivan, even used a quadrature diagram as his office logo. Instead of transcendental symbolism, however, Wright used geometry as a tool to liberate himself from what he regarded as the stifling influence of European architecture, and to create something uniquely American.

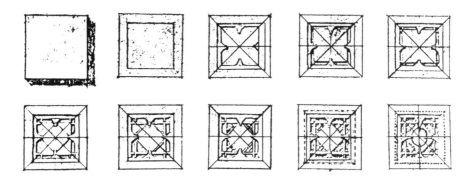

Fig. 13: Louis H. Sullivan, geometrical derivation of organic form

Fig. 14: Frank Lloyd Wright, Unity Temple in Oak Park, Chicago

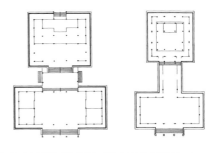

Fig. 15: Frank Lloyd Wright, Unity Temple, ground floor plan

His early masterpiece, the Unity Temple (1906–08), Oak Park, Chicago, USA, is a good example. > Figs. 14 and 15

Historians usually explain the design of Wright's building by identifying earlier buildings and other things that could have served as models. Some claim, for instance, that Wright imitated the cubic style of the German pavilion, designed by Peter Behrens, for the World's Fair in St. Louis, USA, in 1904; other historians say that he adopted a plan typology > Chapter Responses to site, Regionalism from Japanese gongen-style temples, such as the Nikko Taiyu-in-byo. Indeed, Wright's church and these non-Christian temples feature two major masses, one almost square in plan and the other a longer rectangle, connected by a subordinate element. > Fig. 16

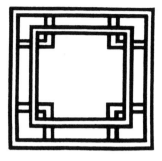

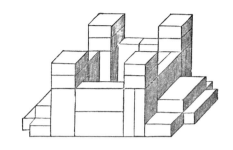

Fig. 16: Arthur Dow, ornament study

Fig. 17: Friedrich W. A. Fröbel, toy house out of woodblocks

Other critics maintain that the design was derived from the compositional principles of Japanese art, as articulated by the painter, Arthur Dow. Wright himself, however, suggested that the Unity Temple was inspired by Fröbel blocks, a toy set with which he played as a child. > Fig. 17

While all of these ideas are to some degree plausible, they explain only some aspects of the design. A geometrical analysis of the building is useful here.

In 1928, Wright published an analytical diagram suggesting that the Temple and adjoining Unity House had been planned using a simple grid with the module of 7 feet. The positioning of windows, skylights, and a few other details certainly match this grid but it is difficult to make the major volumes agree with it. To understand them, one has to reconstruct a different modular grid, based on dividing the central sanctuary into four quadrants. > Fig. 19 Accordingly, the window walls of the temple define a square with 16 such units. The connecting bridge is likewise two units long, as is the central meeting room, if we include the fireplace wall. The sewing room behind it is half a unit deep (a). In fact, the meeting room is based on exactly the same original square as the sanctuary, but in this case, the columns and walls fall within the modular lines.

However, in order to avoid the additive, rigid appearance that often plagues modular plans, Wright also inserts dimensions that have been derived through quadrature and are incompatible with the grid. Hence, for instance, the corner towers in the temple can be derived by taking

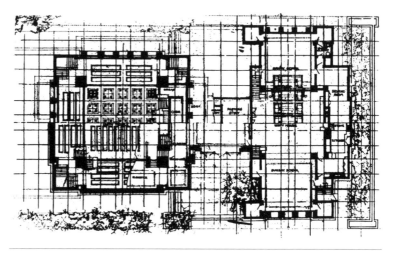

Fig. 18: Frank Lloyd Wright, a gridded plan of Unity Temple

the diagonal of the central space and rotating it by 45° (b). Similarly, the side wings and the front wall of the Unity House correspond to the dimension of a double square, the side length of which is the diagonal of the original square (c). The same operations are used at every scale, even down to ornaments (d). The coexistence of different organizing principles gives the design a particular tension without making it appear random or incomprehensible. > Figs. 18 and 19

Much simpler is the geometry that determines the facade of a small parish church (1966–78) in Riola, near Bologna, Italy. It was perhaps to symbolize the Holy Trinity that Alvar Aalto decided to apply the method of triangulation. The facade of the Riola Church is easily generated using a compass and a 30-60-90° triangle. ABC is a right-angled triangle, such that AC forms the base or ground level, and point B is at an angle of 60° from point A and 30° from point C. By placing a compass at point C it is possible to determine the locations of the clerestory windows that bring natural light into the interior of the church: a line drawn from point B at a right angle to AC will intersect AC at point D, thus determining the first clerestory window. The iteration of the process yields the other four (lines EF, GH, JK), related to each other at the ratio of $\sqrt{3}$. If the line GH is extended upwards to point L, a line CL is at an angle of 60° to the base; drawing the line from L to an extension of the base line at point M completes an equilateral triangle which will determine the southern wall of

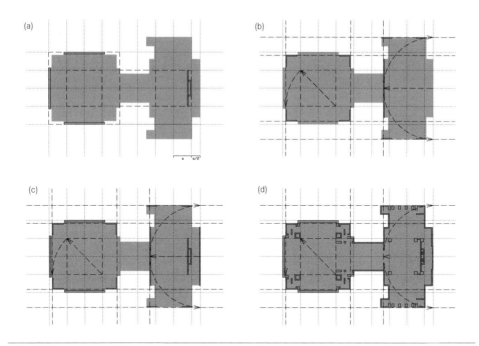

Fig. 19: Frank Lloyd Wright, Unity Temple, geometrical derivation of plan

the church. The angle of the interior concrete arches can be determined by extending the geometrical progression along line CB by two additional steps. Moreover, the curves of the vaults can be drawn by placing the compass at points U, V, X, and Y. > Fig. 20

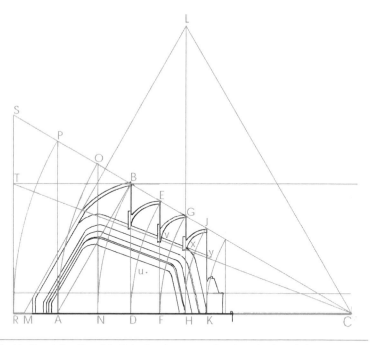

Fig. 20: Alvar Aalto, Church in Riola, facade construction through triangulation

Music and mathematics as models

MUSICAL ANALOGIES

The mathematical structure of Gothic architecture, resulting from the use of quadrature and triangulation, inspired philosophers to describe architecture as frozen music. For many architects, the analogy raised the question whether it would be possible to translate musical compositions and sounds into spatial or architectural configurations.

One way of translating sound into visible form was to use the method devised by German physicist Ernst Chladni in 1787. He spread fine sand over a glass or metal plate, which he made vibrate by moving a violin bow along its edge. Sliding off the points of the most intense vibration, the sand formed complex patterns, depending on the points of support, the position and speed of the bow, the thickness, density and elasticity of the plate, etc. In an analogy to the Chladni figures, Bragdon suggested that architecture is merely the fixed materialization of ephemeral sound patterns. > Figs. 21 and 22

Another way of deriving architecture from music involved translating the tone intervals in a melody into numbers, which would then be interpreted as a spatial system. This investigation led Claude Bragdon to experiments with "magic squares" or matrices in which for example the numbers in each column and row add up to the same sum. He drew a line from one cell to another in the numerical order, creating an accidental, complex figure. > Fig. 23

The two leading artists in the Bauhaus in Weimar, Germany, Wassily Kandinsky and Paul Klee, developed different ways of turning a musical idea into something visual. In accord with his general theory of points, lines and planes, Kandinsky presented in 1925 an alternative to the traditional notation used in music. However, his transcription of the first bars of Beethoven's *Fifth Symphony* represents the notes much like the traditional notation, although the staff is not shown. Also, as in traditional notation, Kandinsky's is read from left to right, the pitch of a note corresponds to how high it is positioned relative to other notes, and the horizontal distance between notes indicates their duration. The size of the points indicates the dynamics. > Figs. 24 and 25

Fig. 21: Ernst Chladni, Chladni patterns of sand on a vibrating plate

Fig. 22: Ernst Chladni, vibrating Chladni plate

Fig. 23: Claude Bragdon, numeric generation of form

Fig. 24: Wassily Kandinsky, notation for Beethoven's Fifth Symphony

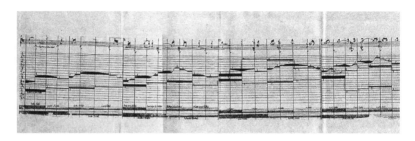

Fig. 25: Paul Klee, Bach Adagio #6

By contrast, Klee's 1924 transcription of the Bach *Adagio No. 6 from the sonata for violin and harpsichord in G major* is more rigorous: the traditional staves have been replaced by a unified grid of parallel horizontal lines, the pitch of a note corresponds to its height on the grid, while its duration corresponds to its length, and its dynamics are represented by the variable thickness of the line.

In 1991, architect Steven Holl used this transcription method to design the facade of the Stretto House in Dallas, Texas. He argues that architecture's expression of mass and materials according to gravity, weight, bearing, tension, and torsion reveal themselves like the orchestration of musical compositions. In the Stretto House, Holl responds to Béla Bartók's *Music for Strings, Percussion and Celeste* of 1936, notable for its strictly symmetrical fugue construction as well as the spatial plan

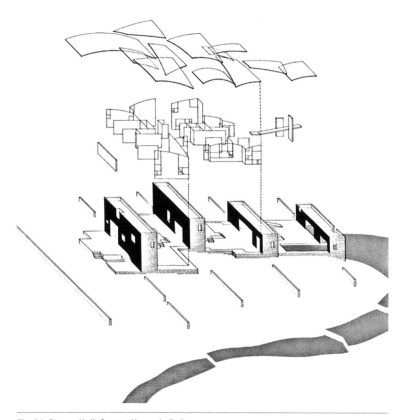

Fig. 26: Steven Holl, Stretto House in Dallas

Fig. 27: J. S. Bach, fugue in E minor, #8 in the Well-tempered Clavier I, BWV 859, measures #52–55

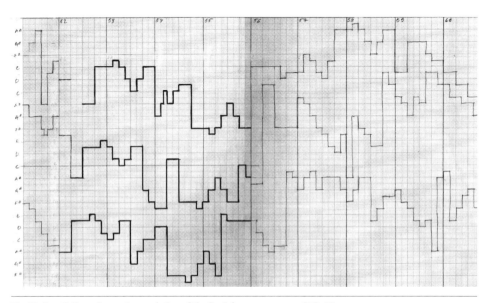

Fig. 28: Henrik Neugeboren, transcription of the Bach fugue, measures #52–55

for the orchestra, which advises that the percussion instruments, celeste, piano, harp and xylophone should be placed in the middle of the stage, flanked on both sides by a string quartet and a double bass. Analogous to this spatial division of the instruments into "heavy" and "light", Holl also contrasts four heavy concrete blocks with curved lightweight roof elements. > Fig. 26

The Bach Monument (1928), designed by Henrik Neugeboren, also follows in the footsteps of Klee but takes one step further. The work is a quasi-three-dimensional notation of a small fragment (measures #52–55) of Bach's fugue in E flat minor in the *Well-tempered Clavier, Book I*. In Neugeboren's system, the x-axis represents the duration and

Fig. 29: Henrik Neugeboren, monument for Bach

the y-axis the pitch of the notes. While in traditional notation the precise duration of each note is represented by a specific symbol so that the actual position of the note on paper does not really matter, in Neugeboren's version each measure is equally wide and divided into regular intervals. Presumably, the tempo should determine the width of the time segments. The most radical departure from standard notation, however, is the vertical z-dimension, which in effect duplicates the information already contained in the y-dimension: the higher the pitch, the higher the corresponding element is positioned along both axes. Neugeboren's decision to represent each voice as a continuous line, which in the three-dimensional model is folded out of a continuous sheet of cardboard: only the segments that are parallel to the x-axis represent notes, while the perpendicular segments are structural necessities without any musical meaning. > Figs. 27–29

While both Holl and Neugeboren developed a three-dimensional form from musical notation, many architects have also proposed structural analogies. Bach's fugues are based on a generating theme or "subject", which is imitated and varied by two or three other contrapuntal voices, called "answers" and "countersubjects". In some Baroque fugues, the variations are strictly regulated: the variation may be a transposition of the subject, an inversion of it, a retrograde version, or even a retrograde inversion. A similar system was applied in the 20th century by Arnold Schönberg and Anton Webern in their 12-tone compositions, in which the same row of twelve notes is played straight, backwards, upside down and upside down backwards. Peter Eisenman's early house designs are as strictly defined as compositional procedures where the initial shape is transformed step by step until the desired complexity is achieved.

HIGHER DIMENSIONS

Dissatisfied with traditional translations of music into architecture, Bragdon went on to construct an instrument that translated a musical score into moving compositions of colors projected onto a screen by electric light. With this "Luxorgan" he even composed and performed massive spectacles, such as the "Cathedral Without Walls" in Central Park, New York, in 1916. Five years later, musician Thomas Wilfred built an instrument called the "Clavilux", also known as the color organ, where Bragdon's "four-dimensional designs" would be translated into motion by using transparent material sensitive to sound, that is, a modernized version of the Chladni plates. Up until the 1950s, Wilfred constructed a number of different Claviluxes, some of which projected light patterns onto walls while others were comprised of a 2-foot square screen in a decorative cabinet resembling a television set. > Fig. 30

The core mechanism of Wilfred's Clavilux consisted of one or more light sources, several gem-encrusted disks, and distorting mirrors revolving at various speeds. Even though some of the light patterns were repeated in cycles, the *lumia* were consistent with Bragdon's principles of four-dimensionality, as no single arrested image could be taken to represent the whole composition.

Bragdon was one of many architects fascinated by the notion of four dimensions, even before Einstein's theory of relativity made specific use of a four-dimensional geometry, with time understood to be the fourth dimension. Inspired by other theosophists, Bragdon postulated the existence of an invisible archetypal world of four dimensions, which we can intuit only imperfectly. > Fig. 31

Towards the end of the 20th century, many avant-garde designers returned to this theme. Peter Eisenman's design for a science building at Carnegie Mellon University, Pittsburgh, USA, was derived from a series of Boolean cubes intersecting each other; distorted solids and open steel frames recorded a trace of these geometries. > Fig. 32

With virtual reality techniques it is possible to apply four-dimensional geometry more consistently. An example is provided by the "invisible architecture" that Marcos Novak exhibited at the Venice Biennale of 2000.

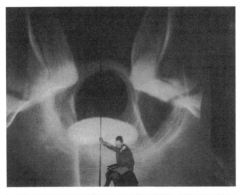

Fig. 30: Thomas Wilfred, Clavilux

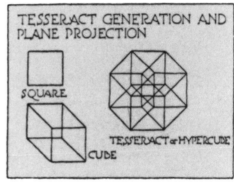

Fig. 31: Claude Bragdon, four-dimensional hypercube

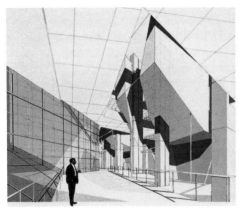

Fig. 32: Peter Eisenman, Research Institute, Carnegie Mellon University (project)

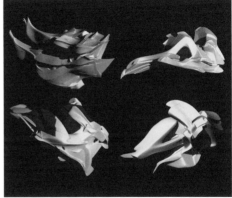

Fig. 33: Marcos Novak, four sculptures as the projections of a four-dimensional object

The main part of the installation was a bar, suspended at each end from wires, with an infrared sensor and lens apparatus. This apparatus created a distinct three-dimensional shape that was invisible to the human eye but could be monitored by a computer. When the hand of a viewer reached into this invisible shape, the computer measured the intrusion and responded with a specific sound. Thus, viewers could divine the location and exact shape of the invisible sculptures by moving their hands and attending to the changing soundscape. > Fig. 33

From a historical perspective, Novak's installation can be seen as combining two inventions popular in the 1920s: the light shows of Bragdon and Wilfred, based on four-dimensional calculations, and another short-lived fad, the Theremin, a musical instrument invented by Léon Theremin in 1919, which translates the motions of a human body electronically into sounds.

PROPORTIONS

A traditional way of working with music involves the use of proportional theory. Pythagoras had already found that the basic intervals in a musical scale can be described using the simple numerical ratios of the holy *tetraktys*, 1+2+3+4. Generalizing from this, the Pythagoreans postulated that the entire universe is built according to geometrical and numerical specifications that harmonize with each other on every scale.

The historian George Hersey has argued that such Pythagorean ideas were developed further by the Neoplatonic philosophers and architects of the Renaissance. He goes as far as to claim that Renaissance architects thought in terms of a Neoplatonic hierarchy, where the physical building was virtually without value, being merely a larger version of the scale model, which was only an imperfect rendering of the design drawings, which in turn were but a shadow of the real thing, the abstract geometrical structure that exemplified the workings of the universe.

Rather less dramatically, historian Rudolf Wittkower suggested that Palladio, one of the geniuses of the Renaissance, designed his villas in such a way that the proportions of the rooms would form a harmonious sequence comparable to a musical fugue. In 1947 Wittkower also presented a seminal interpretation of Palladio's villas, claiming that although they look different from each other, they are all based on a similar diagram, a tartan or band grid.

Wittkower's brilliant essay became famous and inspired theoretician Colin Rowe to discover a similar grid in Le Corbusier's Villa Stein at Garches; he later pointed out similarities between Palladio's Villa Rotonda in Vicenza and Le Corbusier's Villa Savoye at Poissy. > Fig. 34 The suggestion was originally made on the basis of formal similarities only; later historians, however, have unearthed evidence showing that Le Corbusier had actually had an early interest in the work of Palladio. In his influential 1923 book *Towards a New Architecture,* Le Corbusier explicitly advocates the use of geometrical "regulating lines" and a few proportions, including the Golden Section. As an example, we can take a look at the studio he designed in 1922 for his Purist colleague Amédée Ozenfant in Paris.

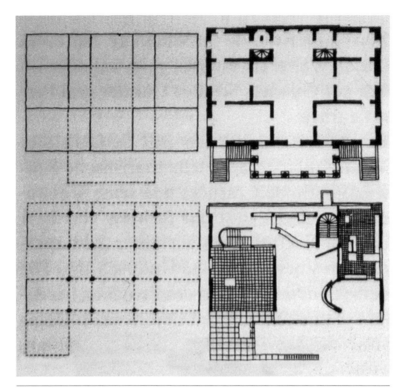

Fig. 34: Colin Rowe, plan grid of Villa La Malcontenta (Palladio 1560) and Villa Stein (Le Corbusier 1927)

In the atelier building, Le Corbusier started with the contingencies of a given, irregularly shaped site. > Fig. 36 Instead of being rectangular, one side of the site is bound by a wall that happens to be at an angle of 30° to the other. The plan is a negotiation of two grids at an angle of 30°. The position of the external stair is determined by continuing the line of the 30° side wall and drawing a perpendicular to this axis so that it hits the front right-hand corner of the site (a). Another stair in the interior is positioned on the intersection of two lines: one connects the wall behind the external stair with the inner right-hand corner of the site, the other is a tangent of the external stair, parallel to the left-hand side wall; the latter line also determines the position of a window in the back wall and the edge of the large atelier window on the side facade (b). The same point is also defined by a line drawn perpendicular to the left wall, from the center of the inner stair to the right-hand wall (c). The edge of the atelier window on the side facade can be determined by drawing a line parallel

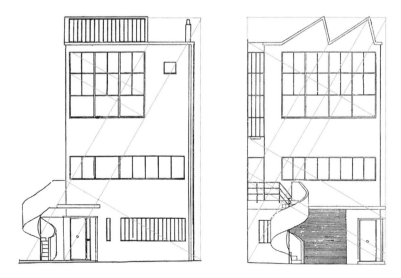

Fig. 35: Le Corbusier, hidden regulating lines in the facade of Ozenfant Studio

to the right wall, from the back window down. Even details are derived from these regulating lines: for example, the corner of the room with the sink, the "laboratory", is on the line going to the left corner of the main facade (d). The same angles govern the facade. > Figs. 35 and 36

Later in his career, Le Corbusier tried to systematize the proportional method in his Modulor system (originally published in 1948). He was convinced that the Golden Section was the key to beauty, but as an irrational proportion it is not easily applicable to building construction, especially with industrial prefabrication. In order to define practical dimensions, he applied the principle of the Fibonacci series, a sequence of numbers generated by adding together two terms to yield a third (1, 1, 2, 3, 5, 8, 13, etc.), so that the ratio between any two terms progressively approximates more closely to the golden section. The Modulor consists of two Fibonacci series, the red one culminating at 183 cm and the blue one at 226 cm, representing for Le Corbusier the height of an ideal man with his arm raised.

For skeptics who feared that the system would limit the creativity of architects and tend towards uniform box-like buildings, Le Corbusier applied the Modulor in two of his most eccentric designs, the Notre-Dame

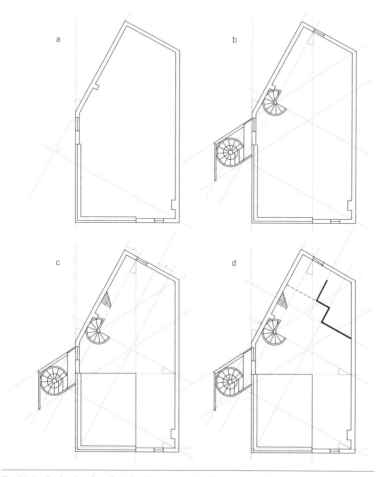

Fig. 36: Le Corbusier, Ozenfant Studio, generative diagram showing design process

du Haut chapel (1954) in Ronchamp, France, and the Philips Pavilion (1958) in Brussels. > Fig. 37

Even if we did not share Le Corbusier's belief in the necessary beauty of the Golden Section, it can be argued that use of consistent proportions throughout the design will make it easy for the viewer to establish visual connections between different elements in the composition. As a result, the design becomes a kind of text, the deciphering of which can be interesting.

Fig. 37: Le Corbusier, Iannis Xenakis, the free Form of the Philips Pavillion is based on ruled surfaces in modulor dimensions.

Another reason for the use of proportional systems is not aesthetic but practical. Inspired perhaps by the Japanese system of tatami mats, many modern architects have adopted proportional systems that work with modules. They allow for the combination of prefabricated elements and standardized components.

Accident and the unconscious as sources

HETEROTOPIA

Although Alvar Aalto, like most architects of his generation, occasionally applied proportional systems to determine the building design in detail, he is usually pictured as the champion of an anti-methodical approach to architecture. His early masterpiece, the Villa Mairea (1939) in Noormarkku, Finland, is said to embody such principles as "forest space" or "Cubist collage." The point is that the rich palette of materials and forms in the building are not tied together by any concept but only by a sensuous atmosphere. A little more analytically, architect and theorist Demetri Porphyrios argued that Aalto's architecture has a particular generating order, namely "heterotopia." However, although Porphyrios claims that in heterotopia there is no organizing principle that would collect the different forms together, other writers have suggested that Aalto takes the path of a visitor who enters the building as the original organization around which other functions are arranged and distorted when it makes the circulation more fluid. Another aspect of Aalto's heterotopic designs is that the variety of forms and organizations is used to highlight the spaces that house the most important functions. Thus, he often celebrates the public spaces with an unusual shape, most notably a fan-shape that resembles a Greek theater plan, and arranges the banal functions (depending on the building, these might be offices, technical spaces, standard apartments, etc.) in a repetitive, simple pattern.

Aalto's design for the Cultural Center in Wolfsburg (1962) exemplifies the heterotopic method. Both the plan and the facade's major programmatic elements have their specific organizational and aesthetic principles. The polygonal auditoria above the main entrance are arranged in a fan pattern and marked in the exterior walls with a striped marble cladding. The offices, by contrast, are placed in an orthogonal order behind a modernist facade that looks like an elongated version of the Villa Savoye by Le Corbusier. Other relatively independent motifs are also introduced, such as a Roman atrium house motif with a tent-like roof and an open fireplace. Instead of a unifying grid, such as we might find in a Mies van der Rohe building, > Precedent, Transformation of a specific model Aalto first gives each programmatic element its own identity and specific form and then packs the elements so close together that their ideal forms are distorted.
> Figs. 38 and 39

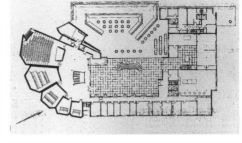

Fig. 38: Alvar Aalto, cultural center Wolfsburg

Fig. 39: Alvar Aalto, cultural center Wolfsburg, 1st floor

While critics have debated the notion of heterotopia, Aalto himself has suggested even stranger ideas about his design method. In one essay, he claimed that he tries to ignore most of the information pertinent to the design and draw almost child-like scribbles; and in another connection, he described his design approach as "play." While the main building of the Experimental House in Muuratsalo (1953), Finland, can be seen as a romantic ruination of a more sober Roman atrium house, the strange (only partly realized) "tail" section looks different. Indeed, it has been suggested that the Experimental House involved playing with drawings of different kinds and in different scales, using landscape features in miniature for free-form pavilion plans, and even mixing portraits with site plans. One of Aalto's recurrent devices is to use the same formal motifs at radically different scales: thus, one of his signature motifs, a fan organization, appears as the leg joint of a stool, as the ceiling of his church in Wolfsburg, as the plan of Seinäjoki library, and the site plan for a housing development in Kotka, Finland. > Figs. 40–42

SURREALIST DEVICES

Some contemporaries of Aalto more precisely articulated anarchic design methods. For instance, Josef Frank promoted a notion of "accidentism," which involved the quasi-accidental combination of various images both from high culture and lowbrow kitsch, in order to achieve the kind of vitality that characterizes naturally grown cities. > Fig. 43

The notion of accident as the driving force of artistic or architectural creativity has been very popular in the 20th century. The roots of the idea are found, however, in Antiquity: Aristotle comments on figures seen in clouds and Pliny tells of Protogenes who created paintings by throwing a sponge against the wall. Inspired by these classics, Leonardo remarked that a wall spotted with stains contains landscapes, battles, and faces.

Fig. 40: Alvar Aalto, weekend house Muuratsalo, site plan

Fig. 41: Alvar Aalto, plan sketch

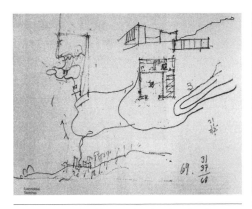

Fig. 42: Alvar Aalto, Muuratsalo, site plan sketch

Fig. 43: Josef Frank, accidentist architecture

His comments were developed into a veritable theory of aleatorism in 1785 by Alexander Cozens, an English landscape painter whose treatise *A New Method of Assisting the Invention in Drawing Original Compositions of Landscape* describes "a mechanical method ... to draw forth the ideas" of artists. It consists of making casual and accidental ink blots on paper with a brush; the paper may be first crumpled up in the hand and then flattened. Cozens stresses that the blot is not a drawing, but an assemblage of accidental shapes, from which a drawing may be made. After selecting a suggestive blot, the artist should trace an image out of it without adding anything not in the blot. The drawing is finished by adding ink washes. The purpose of this method was to free the artist from involuntary servitude to conventional schemes of landscape composition by relinquishing deliberate control. > Fig. 44

Fig. 44: Alexander Cozens, inkblots

Similar methods were applied in the 20th century by the surrealists, who also revived the spiritist and theosophist technique of automatic writing. While the theosophists had thought of automatic writing as a situation in which the medium releases his or her self in order to let spirits guide the hand, the surrealists reframed the practice in psychoanalytic terms. Adopting a popular parlor game, the first generation of surrealist painters toyed with the method of "exquisite corpse" in order to replace an individual author by a group. The first person draws something on top of a sheet of paper and folds the paper to conceal the drawing, except for a few points to which the next person has to connect. Max Ernst's favorite method was *frottage,* creating automatic images and patterns by making a rubbing on a piece of paper placed on top of a textured surface. He also applied *grattage,* a technique in which paint is scraped off a canvas.

Later surrealists came up with several additional techniques to produce images. In Romanian artist and poet Gherasim Luca's *cubomania,* several images are cut into identically sized squares, which are then rearranged at random to make surprising new combinations. *Soufflage* is a technique in which liquid paint is blown onto a surface; in *parsemage* charcoal dust is scattered on the surface of water and then skimmed off with a sheet of paper; in *fumage* images are imprinted on a sheet of paper or canvas by the smoke of a candle or a kerosene lamp; *entoptic graphomania,* a variation of automatic drawing as devised by Luca's compatriot artist Dolfi Trost, consists of marking accidental impurities in the paper with dots that are then connected with lines. One of the few three-dimensional methods is *coulage,* a technique in which molten wax, chocolate or tin is poured into cold water to create an accidental sculpture.

Later, Trost went on to abandon such artistic techniques of surrealist automatism in favor of those "resulting from rigorously applied scientific procedures." However, in the latter cases as well, the result would be unpredictable.

There are also a few architects who apply surrealist methods to determine the forms of their buildings. Wolf Prix and Helmut Swiczinsky of Coop Himmelb(l)au revived automatic drawing in their Open House project for Malibu, California (1990). The design was created "from an explosive-like sketch drawn with eyes closed. Undistracted concentration. The hand as a seismograph of those feelings created by space." While one of the architects was drawing, the other converted the sketch into a three-dimensional model without censorship or evaluation – while *Purple Haze* by Jimi Hendrix would blast out of high-performance loud-speakers. > Figs. 45 and 46

Many other contemporary architects have been influenced by Surrealism, even if surrealist methods have not been used directly for form-giving. Rem Koolhaas, who explicitly referred to Salvador Dalí's Paranoid-Critical Method, has focused on programmatic effects rather than formal configurations, and has recommended overlaying incompatible programs into a discontinuous whole that is expected to engender new events. One example is the OMA design for the Parc de la Villette (1982) in Paris, which is a montage of incongruent programmatic elements. In more

Fig. 45: Coop Himmelb(l)au, Open House (project), sketch

Fig. 46: Coop Himmelb(l)au, Open House, model

1

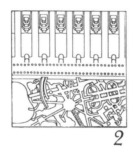

2

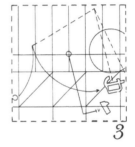

3

Fig. 47: Bernhard Tschumi, Manhattan Transcripts, translating a story into architecture

Fig. 48: R&Sie, Dusty Relief/B-mu, (project) section

Fig. 49: R&Sie, Dusty Relief/B-mu, View of the building

precise terms, architect and theorist Bernard Tschumi divides such operations into dis-, cross-, and transprogramming. He has also explored the potential of montage, with explicit references to film, and proposed elaborate systems of notation. Tschumi's *The Manhattan Transcripts* (1978/94) translates a detective novel narrative into an architectural project. > Fig. 47

A more rigorously surrealist project is the design for a Contemporary Art Museum in Bangkok, Thailand, by R&Sie (2002), entitled Dusty Relief/B-mu. It involves, among other things, a random relief calculated from the pixelization of "aleatory particles for a pure grey ectoplasm"; an electrostatic system that collects the city's dust on the surface of aluminum latticework; and finally, a schizophrenic contrast between the "Euclidean" interior and the "topological" exterior. The result is a building with a facade that constant changes in color, shape and texture in response to air pollution in the city. > Figs. 48 and 49

Rationalist approaches

At the same time as the original surrealists were experimenting with irrational and random techniques, some of the artists and architects in the Bauhaus took the opposite approach and attempted to develop rational and objectively correct methods of design. During his tenure as director (1928–30), Hannes Meyer declared that architecture was not one of the fine arts, and consequently, an architect had no right to act on the basis of subjective intuition or creative inspiration; instead, every architectural design needed to be based on solid, scientific knowledge about that which can be measured, observed or weighed. To build up this knowledge base, he invited a number of scientists to give lectures on new findings in philosophy, physics, economics, sociology, psychology, physiology, anatomy, etc. and urged his colleagues to research construction techniques, materials, and functional organizations. The actual design process would make use of such general knowledge coupled with specific information about the task at hand, in particular, facts about the program and the site. For example, Meyer emphasized the importance of mapping sun angles and measuring the capillary capacity of the soil and the moisture in the air. Once all relevant facts were known, he promised, the architectural design would automatically "calculate itself." Meyer's spirit lives on today in the bubble diagrams that some architects draw to visualize the connections between different functions. > Fig. 50

A project for a community building (1930) by Hannes Meyer, Tibor Weiner, and Philipp Tolziner starts from two diagrams: one describes the sequence in which the functions unfold, the other determines the angles of sunlight. In this case, the movement diagram specifies the following sequence: arrival – change of dress – protected living – unprotected living – and then either back again or change of dress – bathing – change into pajamas – sleeping. The assumption is that different functions have different needs of sunlight: the bedroom should get morning light while the living room should have sunlight in the evening. On this basis, the design arrives at a two-room apartment with bathroom and bedroom functions in the first space and the living room in the second. All rooms are oriented to the south, but the volumes are stacked in such a way that the bathroom will provide shade for the bed in the evening, while the corner window of the living room is turned to the southwest.

Despite such exercises, Meyer never articulated a complete set of instructions for generating a building from the factual information. In fact,

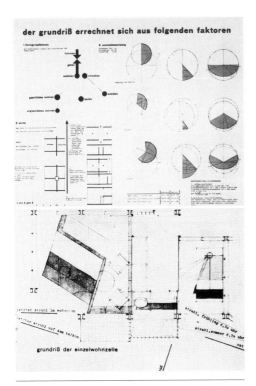

Fig. 50: Study for a scientific derivation of a plan for a housing unit (project)

many of his contemporaries felt that he followed an aesthetic agenda that was not explicitly stated and could not be argued for in objective terms. Hugo Häring, for one, claimed that the likes of Meyer harbored an irrational and unjustifiable aesthetic preference for simple geometrical forms. What Häring himself offered in place of "decadent geometric thinking" was the method of *Leistungsform,* proposing that the shape of a thing should be derived, without prejudice or preconception, from the precise spatial parameters of the desired action.

One famous example of *Leistungsform* is Häring's cowshed (1924–25) at the Gut Garkau in Scharbeutz in the Lake Pönitz district. While a typical shed has a rectangular plan that is easy to construct, connect to other buildings, or expand, Häring's design features an unusual oval configuration. Such a plan is more complicated to build and arguably less flexible to change but, according to the architect, it optimizes the way the cows move in and out of the shed. > Figs. 51 and 52

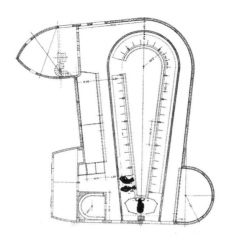

Fig. 51: Hugo Häring, Gut Garkau, ground floor plan as Leistungsform

But how can the correct performance form be discovered? Häring did not promise that there would be a simple procedure to achieve this; instead he spoke of "the secret of the origin of form." Yet his notion of Leistungsform could be linked with work efficiency studies that were popular among the functionalists in the 1920s. While the pioneer of Scientific Management, Frederick Winslow Taylor, only measured the time needed for separate steps in the work process, Frank B. Gilbreth made use of chronophotography and the movie camera to accurately map the worker's movements as luminous white curves against a black background; later he also constructed wire models, called "cyclographs," that would show the optimal trajectories in three dimensions. With Gilbreth's cyclographs, it might be possible to determine the performance forms that Häring wanted to find. > Fig. 53

The optimization of the form is possible when the function is very clearly defined. Thinking about a dwelling, however, we soon realize that most spaces serve a large range of functions, and if the shape of the room is really optimized for any particular function, it is usually less than ideal for many of the other ones. We have to decide what we should optimize: the amount of space, the costs, the convenience of use, all of the above at the same time, or something completely different.

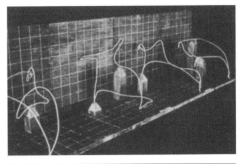

Fig. 52: Hugo Häring, Gut Garkau, cowshed   Fig. 53: Frank B. Gilbreth, motion study

DESIGN RESEARCH

The project of design research gained new momentum in the 1960s with the development of affordable computer applications. Nicholas Negroponte imagined an architecture machine that would produce designs on its own, and George Stiny and William Mitchell developed shape grammars as a way of generating architecture with a computer. Applying Chomskian linguistics to Wittkower's analysis of Palladio's villas, Stiny and Mitchell wrote programs producing new plans and elevations with Palladian characteristics. All of the new designs had elements that characterize the Renaissance originals, including porticos, temple fronts, and undecorated, vernacular blocks; the plans were variations of the irregular tartan grid that Wittkower had discovered in the originals.

The productivity of the computer, however, begs the question of whether there are any good villas among the thousands generated. The project of generating architecture through the computer involves the production of formal solutions of some *relevance;* hence, the generative program without a sorting or evaluating possibility is of little value. However, if we can define the desired parameters to evaluate the variations, then it is possible that we could reach a good solution without going through all the variations.

Bill Hillier and Julienne Hanson's space syntax is another method that attempts something more ambitious than imitating the style of a historical architect. They claim that social relations are irreducibly spatial and vice versa; both are a matter of configurations, either of people or spaces. Although the social and the spatial configurations are called "morphic languages," they do not symbolize or signify anything but themselves. Hillier and Hanson assume two kinds of actors, inhabitants and visitors, whose spatial behavior they try to model. An important factor is the depth

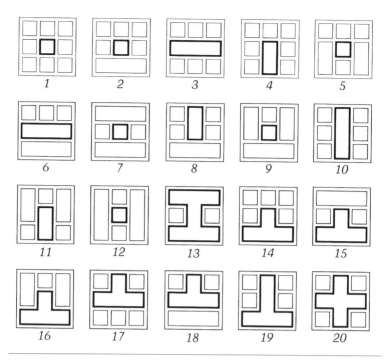

Fig. 54: William Mitchell, all nine-square variations following rules 1–19 of the generative grammar of Palladian villas

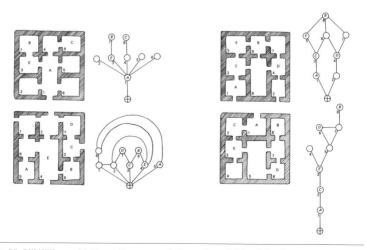

Fig. 55: Bill Hillier and Julienne Hanson, variations of spatial depth in nine-square plans

of the spaces. The deeper the space – i.e. the greater number of rooms or other spaces one has to traverse to enter – the more separated it is, and the higher in status and power. The authors give the example of four plans, which are formally similar in that they are all based on a nine-square grid, but which have radically different spatial structures in terms of their depths. > Figs. 54 and 55

The space syntax method is provocatively reductivist: cultural conventions and aesthetic factors are bracketed out of the analysis. Claiming that we need not consider the architectural experience in all its richness, Hillier promises that simple map analyzes are enough to make empirically verifiable predictions about people's behavior in buildings and urban spaces. While space syntax is not a method that would generate forms, it can be used to evaluate the effects of alternative designs.

Much easier to apply is the pattern language of Christopher Alexander, whose method is rooted in similar, mathematical as well as empirical observations concerning architecture. He distinguishes, for instance, between requirement diagrams, form diagrams, and constructive diagrams. A requirement diagram describes the constraints that are relevant to the situation; a form diagram defines a precise formal organization, ideally with foreseeable functional consequences; and a constructive diagram combines a formal and a functional explanation of the thing to be designed.

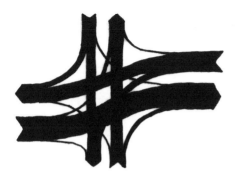

Fig. 56: Christopher Alexander, constructive diagram of a traffic crossing

NR40a
DO:08 08 97

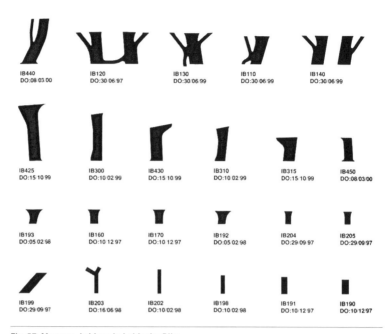

IB440
DO:08 03 00

IB120
DO:30 06 97

IB130
DO:30 06 99

IB110
DO:30 06 99

IB140
DO:30 06 99

IB425
DO:15 10 99

IB300
DO:10 02 99

IB430
DO:15 10 99

IB310
DO:10 02 99

IB315
DO:15 10 99

IB450
DO:08 03 00

IB193
DO:05 02 98

IB160
DO:10 12 97

IB170
DO:10 12 97

IB192
DO:05 02 98

IB204
DO:29 09 97

IB205
DO:29 09 97

IB199
DO:29 09 97

IB203
DO:16 06 98

IB202
DO:10 02 98

IB198
DO:10 02 98

IB191
DO:10 12 97

IB190
DO:10 12 97

Fig. 57: Maxwan, bridges in Leidsche Rijn

One of Alexander's examples of a constructive diagram is a representation of traffic flows at a congested street intersection. Instead of using numbers to indicate the number of cars, Alexander makes a busier street wider than a more quiet one. What emerges is an image of the form that the intersection must take to meet the functional requirements.

Precisely this method was applied by the Dutch office Maxwan (Rients Dijkstra und Rianne Makkink) to design 30 bridges (1997–2000) for the new town of Leidsche Rijn in the Netherlands. Each bridge was tailored to the forecast volume and types of traffic, with separate decks for each type of user. > Figs. 56 and 57

Alexander maintains that one of the major problems in architectural design is the tendency to formulate questions verbally. Instead of working with abstract linguistic concepts, he wants to decompose each design assignment into concrete, partial problems, solve them, and integrate them into a hierarchical whole. On this basis, Alexander developed a pattern language (*A Pattern Language,* 1977), which matches universally valid formal solutions to patterns of events, producing "the quality without a name" that is "shared by good architecture of any time or place." He claims that this quality is made of 15 fundamental components: levels of scale; strong centers; boundaries; alternating repetition; positive space; good shape; local symmetries; deep interlock and ambiguity; contrast; gradients; roughness; echoes; the void; simplicity and inner calm; and finally, non-separateness. These qualities are said to be exemplified by his own design for a column for a building in San Jose, California.

Alexander described building as an endless process in which the best result is achieved when the architect helps the residents to find the right patterns. In this sense, the pattern language can be considered one of the most advanced concepts of user planning: it enables non-architects to conceive of the options and their consequences, and then make informed choices of what should be built.

With patterns, Alexander means relations of objects at any scale, from cities to construction details. The patterns propose a solution type to a common design problem but are not intended to be always reproduced in exactly the same way. In *A Pattern Language* there is a brief verbal description of each pattern, an argument explaining its benefits, and usually a visual diagram. Each pattern must be related to others: the larger-scale ones that include it as well as the smaller-scale ones included in it. However, the method is a hierarchical one: one must choose, according to Alexander, an initial pattern that serves to determine the other ones.

In 1980, Alexander was invited to design a café for a large exhibition building in Linz, Austria. He claims to have applied 53 patterns out of the total of 253 included in the book *A Pattern Language*, but does not specify which pattern is the initial one. Possibly we can take the pattern #88, "Street café" as the starting point. > Fig. 58

This diagram advises that the café should be oriented towards the street as the focus of attention and as a kind of stage. In this case, however, there was no urban street on the site. Instead of a real street, Alexander applies the pattern #101, "passage through a building" to interpret the interior corridor of the long exhibition building as a kind of a street.
> Figs. 59 and 60

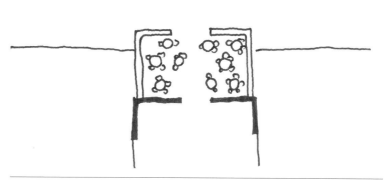

Fig. 58: Christopher Alexander, Pattern #88

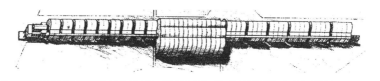

Fig. 59: Christopher Alexander, Linz Café at the end of the exhibition building

The design process, according to Alexander, begins with a consideration of the basic functional pattern as well as the qualities of the site. In Linz, he realized that the café should be oriented towards the afternoon sun and the river, and be high enough to afford views of the river landscape.

He further applies pattern #163 to create a communal place bounded partly with a roof, partly with walls; pattern #161 to orient this space towards the sun; and pattern #176 to provide sitting areas in a green environment.

Alexander felt that the entrance to the café was particularly successful. Applying patterns #110 and #130, he places the entrance so that it is directly visible and accessible from the main thoroughfare, and has a striking form. Moreover, the entry area should be partly in and partly outside of the building.

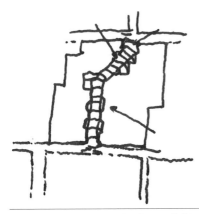

Fig. 60: Christopher Alexander, Pattern #101

As the example of the Linz Café demonstrates, the pattern language leaves many options open, and Alexander himself emphasizes emotional and atmospheric qualities as the ultimate reasons for his decisions. Still, many of the patterns are worth considering, if we keep in mind that they are not universally valid, but rather embody a Mediterranean-Californian ideal of the good life in a sociable, middle-class community. A particular challenge is the task of combining the solutions to partial problems so as to make a harmonious whole.

Precedent

TYPOLOGY

While Alexander claims that one can apply each pattern a million times over without ever doing it the same way twice, some other architects have promoted more strictly defined solutions. At around 1800, Jacques-Nicolas-Louis Durand proposed a typological theory that sees architecture as the art of arranging a given set of elements (columns, entrances, staircases, etc.) in an orthogonal composition in order to find a simple and economical solution. Columns were to be placed at the intersections, walls on the axes, and openings at the centers of the modules. Durand's *formule graphique* can be seen as a precursor for later applications of standardization and prefabrication of building parts.

Most theorists of typology, however, have proposed a different understanding of type, one that allows for a certain amount of variation in the standard elements. In general, typology involves a classification of buildings (or building parts) on the basis of formal or functional similarities. A basilica church, for example, is a type characterized by a linear plan organization with a central nave flanked by two or four side aisles that are lower so that the nave is partly lit by clerestory windows. This typology has been popular for centuries, and there are thousands of churches that look different but nonetheless belong to the basilica type.

In the 1960s, architects such as Aldo Rossi revived typology as a design method. He had a passion for greatly reduced structural and spatial types, evolved from vernacular and classical traditions, which he claimed make sense of the environment and embody the collective memory of the community. Rossi always insisted that types are conceptual constructions and never identical with the physical forms of buildings. Yet he used basic types in a very pure form in his buildings, no matter what the scale or the function. Thus, the shape of an octagonal tower appears in a secondary school in Broni (1970), Italy, in Rossi's famous theater-boat, the Teatro del Mondo (1979), Venice, and in his design for a coffee pot, "La Conica" (1982), for Alessi. The entrance to the school in Broni features a simplified classical temple front with a clock, similar to Rossi's design for a large theater in Genoa (1990), a small changing cabin for a beach, and an exhibition case. > Figs. 61–63

A more traditional understanding of type stresses its flexibility. In the early 19th century, Antoine-Chrysostôme Quatremère de Quincy distinguished between *model* and *type:* "The model is an object that should

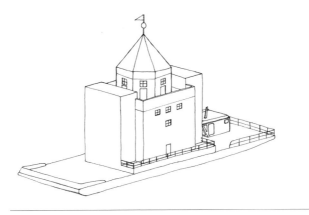

Fig. 61: Aldo Rossi, Teatro del mondo, Venice

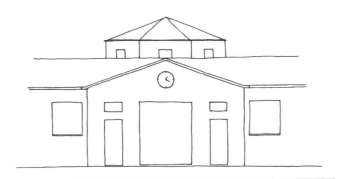

Fig. 62: Aldo Rossi, School in Broni

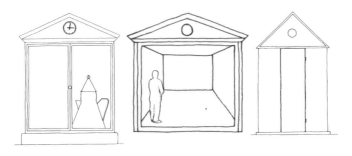

Fig. 63: Aldo Rossi, similar forms in a coffee pot, a display case, a changing cabin and a theater stage

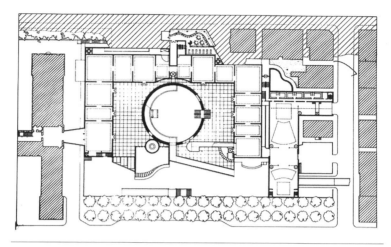

Fig. 64: James Stirling, Neue Staatsgalerie, Stuttgart

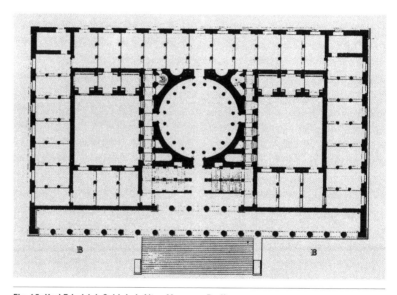

Fig. 65: Karl Friedrich Schinkel, Altes Museum, Berlin

be repeated as it is; the type, on the other hand, is something after which different artists conceive works of art that may have no obvious resemblance. All is exact and defined in the model; everything is more or less vague in the type." Rossi's roster of fixed forms could be seen as models, while the majority of typologically oriented architects are interested

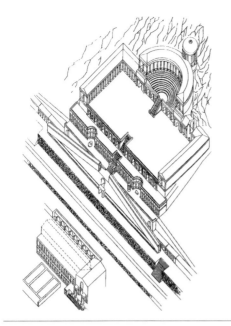

Fig. 66: Sanctuary of Fortuna Primigenia, Palestrina

in the variability of types and their dependence on the historical and social situation.

James Stirling's Neue Staatsgalerie (1978–83) in Stuttgart, Germany is a representative example of postmodern typological design. Instead of striving for a simple, irreducible shape, Stirling combines two independent basic typologies to make an undecidable whole. On the one hand, the plan organization – a rotunda placed inside a rectangular plan – harks back to the museum type established by Karl Friedrich Schinkel with his Altes Museum (1823–30) in Berlin. The row of trees planted along the street front of the Staatsgalerie reproduces the Ionic colonnade of Schinkel's museum. On the other, the way Stirling's building employs ramps to connect to the street suggests another type, evident, for example, in the temple of Fortuna Primigenia (c. 80 BC) in Palestrina. > Figs. 64–66

The principle of combining apparently contradictory typologies has always been popular. The Hagia Sophia in Istanbul is at once a basilica, a cruciform church, and a centralized reinterpretation of the Pantheon. Likewise, Balthasar Neumann's pilgrimage church *Vierzehnheiligen* (1743–72) near Lichtenfels, Germany, combines a longitudinal and a

Fig. 67: Le Corbusier, church in Ronchamp

centralized plan: from the entrance door, the church appears to be a basilica with side aisles, but as one proceeds further inside, the colonnades swing back to suggest a centralized space with the saints' altar in the middle. The Notre-Dame du Haut chapel (1954) in Ronchamp, France, by Le Corbusier does likewise: one side of the church promises a longitudinal organization and the other a cruciform plan. > Fig. 67

Robert Venturi's first building, his Mother's House (1962) in Chestnut Hill, Pennsylvania, could be examined as a deliberately contradictory composition. The design process has been well documented, and it transpires that the architect was not averse to trying out one idea after another, creating ten radically different designs until finally arriving at the version that was built.

The facade combines a generic image of a house with highbrow references to Egyptian pylons, Baroque portals and modernist ribbon windows. > Fig. 68 The main facade is symmetrical, with what appears to be a pitched roof and a giant chimney, as one might expect to see in a child's drawing of a house. However, on one side of the facade we see a semi-traditional square window and on the other, a modernist ribbon window that would not be out of place in Le Corbusier's Villa Savoye. The arch over the entrance suggests that there might be a hidden affinity, though: both sides have five window squares, although on the right-hand side the windows are arranged in a line, and on the left they make two square configurations, one with four, the other with just one window. The arch touches the corner of the solitary window, indicating an absent square of the same size on the right-hand side (d).

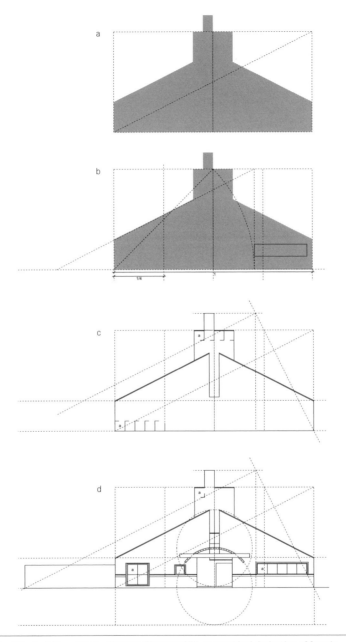

Fig. 68: Robert Venturi, Vanna Venturi House, Chestnut Hill, geometrical derivation of facade

In this case, the typological and morphological references are also controlled geometrically. The facade can be inscribed in a double square, defined by the side walls of the house and the bulk of the chimney (a). If we rotate the diagonal of each square down to make a $\sqrt{2}$ rectangle, we get the position of the windows (b). The roofline can be derived by first drawing the diagonal of the double square and then sliding this diagonal to the $\sqrt{2}$ rectangle edge. (c). > Fig. 68

TRANSFORMATION OF A SPECIFIC MODEL

Typological design is not the only way to respond to precedents. Architects can also make use of specific historical buildings as starting point for the design. Ludwig Mies van der Rohe's design for the German Pavilion in the Barcelona International Exhibition of 1929, for example, owes a lot to a number of precedents, both architectural and artistic. Critics have pointed out that Miesian plans – with walls laid out on an orthogonal grid as independent elements that occasionally project out in the distance – resembled paintings by the De Stijl group. Some characteristics of the Barcelona Pavilion, such as the open glass corners, could also be related to the Prairie Houses of Frank Lloyd Wright. Perhaps more important, however, is the classical tradition. It is no accident that the plan of the Pavilion (as well as the pool next to it) repeats the proportions of the Parthenon in Athens. Moreover, the columns of the temple match the walls of the exhibition building, and the walls of the cella in the Parthenon match the columns in Mies's design. Furthermore, the pool inside the pavilion follows the division in the temple between the *adyton* and the *opisthodomos*. The exhibition building also contained a sculpture of a woman (by Georg Kolbe, 1929), which takes the place of the statue of Athena Parthenos in the Greek temple. These classical overtones were more obvious during the exhibition in 1929 than they are today, as one would view the facade of Mies's building through a row of freestanding Greek columns. > Fig. 69

When using a precedent, it is important to transform it, instead of just imitating its familiar aspects. As another use of a precedent, we can consider the Villa dall'Ava (1991) at St. Cloud near Paris by Rem Koolhaas and OMA. > Fig. 70 Like the earlier villa, Koolhaas' building exemplifies Le Corbusier's *five points* towards a new architecture: pilotis, free plan, free facade, ribbon windows and a roof garden. However, in Koolhaas' building the elements are reshuffled as fragments. Thus, a part of the curved class wall of the ground floor level at the Villa Savoye reappears as the interior kitchen wall in the Villa dall'Ava. Le Corbusier's villa is almost a square while Koolhaas' variation is based on a Golden Section grid, subdivided four times – following the Corbusian principle of regulating lines –

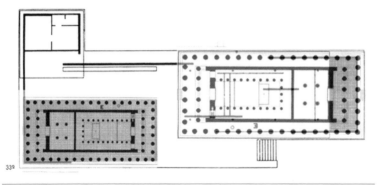

Fig. 69: Ludwig Mies van der Rohe, Barcelona Pavillon with plans of the Parthenon overlaid

in order to determine the dimensions of the cantilevered wings. At the same time, Koolhaas transforms the quoted elements: the roof garden turns into a swimming pool, the abstractly geometrical stucco gives way to corrugated metal facades, the pilotis under the eastern bedroom violate the structural logic of Le Corbusier's Maison Dom-ino (1914). Transforming the Villa Savoye makes sense, for, as mentioned earlier, Le Corbusier's villa was in many ways a similar transformation of an earlier precedent, Palladio's Villa Rotonda in Vicenza. The Maison Lemoine (1998) at Floirac, near Bordeaux, by OMA, can also be seen as a variation on the Villa Savoye. While in the Villa Savoye Le Corbusier reversed the Renaissance principle of a solid wall at ground level and freestanding columns at the *piano nobile* upper story, Koolhaas inverts the Corbusian design in his Bordeaux house: the ground floor is underground, the *piano nobile* is all-glass, and the classical *corona aedificii* is a solid-looking block that hovers in the air. With the Villa dall'Ava the precedent is transformed in multiple ways – some transformations more abstracted, others more obvious direct quotations of the precedent – the resulting design suggests a historical depth. > Fig. 70

The precedent does not have to be an established masterpiece of architecture. The Casa Malaparte (1940) on the island of Capri by Adalberto Libera and Curzio Malaparte is partly based on a precedent: the unusual truncated stair recalls that of a rural church on the island of Lipari where Malaparte had been exiled. > Figs. 71 and 72

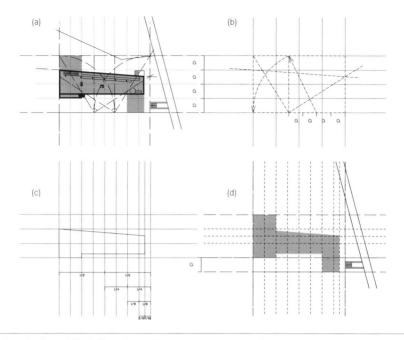

Fig. 70: Rem Koolhaas, Villa dall'Ava, Paris, diagrammatic generation of plan

Fig. 71: Adalberto Libera, Curzio Malaparte, Casa Malaparte, Capri

Fig. 72: Church of Lipari

Architects Herzog & de Meuron often reuse found elements that have no particular architectural value. Thus, for the facade of a housing block on the Schützenmattstrasse in Basel, Switzerland (1993), they took inspiration from the design of manhole covers in the same city, stretching and enlarging the motif, and for the Dominus Winery (1998) in Napa Valley, USA, they applied a retaining wall system common in the Alps.

Responses to site

REGIONALISM

In regionalism, the architect tries to adopt some features not necessarily from the immediate environment but from the region or the nation. Thus, when architect Adolf Loos built the Khuner House (1929) in the mountains in Payerbach, Austria, he designed walls to be built in dark wood, even though in urban contexts he would almost always opt for white stucco. > Fig. 73

In regionalist architecture, we pay attention to local materials and construction techniques, not only for aesthetic reasons, but because local traditions often incorporate tried-and-tested solutions that are sustainable and optimized for the climate, lighting conditions, temperature, humidity, and so on. Hassan Fathy was one of the pioneers of regionalism: in his buildings for the village of New Gourna (1948) in Egypt he applied ancient construction methods such as mud brick (adobe), and used dense brick walls and traditional courtyards to provide passive cooling. In addition to such technical advantages, regionalist techniques also made it possible for Fathy to incorporate local builders, including the inhabitants, in the design and construction process, and to accomplish impressive results at low costs. Villa Eila (1995) in the town of Mali, Guinea, by Finnish architects Mikko Heikkinen and Markku Komonen, is a contemporary example of regionalism, reacting to the strong light and humid air of the tropical West African coast. > Figs. 74 and 75

While Loos's Khuner House resembles traditional architecture in terms of materiality, it lacks some traditional features, such as a high

Fig. 73: Adolf Loos, House in Payerbach

Fig. 74: Hassan Fathy, adobe construction   Fig. 75: Heikkinen and Kommonen, Villa Eila

roof. In this sense, it could also be described as what Alexander Tzonis, Liane Lefaivre and Kenneth Frampton have called "critical regionalism." Frampton assumes an *arrière-garde* position, distancing himself equally from the pre-industrial past and the Enlightenment myth of progress. In his analysis, critical regionalism focuses on local specificities in order to resist the uniformity of capitalist modernity. It deconstructs the overall spectrum of world culture it has inherited and critiques universal civilization. On a more concrete level, Frampton advises architects to use local materials in a tectonic way, i.e. revealing the actual construction, instead of the abstract and generic constructions typical of international modernism.

The houses that Mario Botta constructed in Ticino, Switzerland, during the late 1970s and 1980s can be considered as examples of critical regionalism. As a student of Le Corbusier, Botta applies the simple and abstract geometry of modernism but connects to local building traditions by the use of colors and materials. Thus, many of his houses imitate the striped walls typical of the region.

One of the key examples of critical regionalism as defined by Frampton is the Bagsvaerd Church, near Copenhagen, designed by Jørn Utzon in 1976. With the building Frampton illustrates the notion of a self-conscious synthesis between universal civilization and world culture. The precast concrete element exterior of the building seems like an example of rational universal civilization, with echoes of rural utility buildings, such as grain silos, while the interior, with its organic in-situ concrete shell vault, alludes not only to Western tectonic canons but also to a precedent from the East, the Chinese pagoda roof. As a result, Utzon is able to eschew sentimental *Heimatstil* as well as contemporary ecclesiastical kitsch, and constitute a regionally articulated basis for the spiritual in a secular age. > Figs. 76 and 77

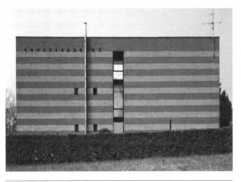

Fig. 76: Mario Botta, House in Ligornetto   Fig. 77: Traditional architecture in Ticino

CONTEXTUALISM

Regionalism is not the only way to respond to the site. Postmodernist architects, such as O. M. Ungers, would draw abstract diagrams about the morphology (roof angles, window axes, texture, etc.) of the environment, and then attempt to reconstruct a new composition with similar characteristics, as in his 1978 project for a town hall in Hildesheim, Germany.

Hans Hollein's Media Tower (1994–2001) in Vienna is an even more direct example of contextualism, as opposed to typological design, in that it does not respect the Viennese block type but collages together an idiosyncratic mixture of elements. With the exception of a tall glass tower, the facades of the building imitate neighboring buildings with regard to colors, proportions, and fenestration. As well as varying the skin to match the surroundings, Hollein also pulls out a peculiar leaning glass box to act as a center of attention at the street corner. > Figs. 78 and 79

In a more abstract way, Richard Meier derived his design for the Arts and Crafts Museum (1980–84) > Fig. 81 in Frankfurt from the site (a). The starting point was a 19th-century villa, which Meier used as a module to define a 4 × 4 grid. From this grid, he then isolated the corner squares to establish a *castello* type, with the old building as one of four corner "towers." The corners suggest two main axes, which Meier defined as walkways, forming a four-square (b). However, the organization is richer and more ambiguous (d).

The new building also forms an L shape framing the old villa as a special piece (c). This L-configuration can be seen as the reason for pushing the two main axes southwest from the center (e). There are, however,

Fig. 78: Hans Hollein, Media Tower Vienna

Fig. 79: Hans Hollein, Media Tower correspondences to adjacent facades

other special elements as well, competing for attention. One is the court-yard that terminates the entry axis. While the interior of the colonnade has the width of the module, it is surrounded by ambulatories that make the open space wider. The width of the courtyard is one-third of the width of the building, suggesting that the open space could be a fragment of a nine-square configuration, with a void as the center of the complex. The third special element is made of the round fragments that form the entrance to the museum.

Pulled together, they define a circle that fits into a square of the original grid (f). The fragmentation results from the contextual inflections Meier introduces. Continuing the orientation of the Schaumainkai street to the west, he rotates the grid by 3.5° around a point marked with a gate at the eastern side of the complex. In response to another bend in the river to the east, Meier also rotates the plan in the opposite direction.

> Figs. 80 and 81

Daniel Libeskind's Jewish Museum (1989–2001) in Berlin is an example of a deconstructivist reaction to a context. Although the architect had used similar, lightning-stroke figures in his earlier designs, and might

Fig. 80: Richard Meier, Arts-and-Crafts Museum in Frankfurt, elevation

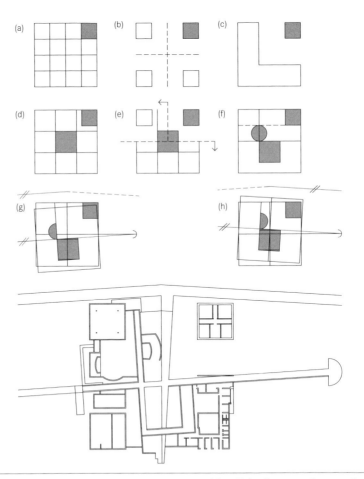

Fig. 81: Richard Meier, Arts-and-Crafts Museum, Frankfurt/Main, diagrammatic generation of the plan

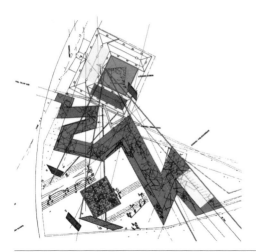

Fig. 82: Daniel Libeskind, Jewish Museum Berlin plan diagram

have been originally inspired by Michael Heizer's land art project *Rift,* the first of his *Nine Nevada Depressions* (1968), here the complicated shape responds to the immediate surroundings, in particular the old Baroque museum of which the new museum was first seen as an extension.

Baroque art often emphasizes diagonals and radial organizations. Accordingly, Libeskind builds his meandering building so that many of the side walls converge towards the center of the original building's back facade. The different widths of the new museum are also taken from the wings of the old, with or without corridors, and the E. T. A. Hoffmann garden in Libeskind's design has the exact dimensions of the old building's courtyard. The composition also includes an axis that cuts across the new museum at several points, creating inaccessible voids; the segments of the axis that are not inside the museum appear as displaced, free-standing solids. > Fig. 82

Generative processes

SUPERPOSITION AND SCALING

Beginning with an exhibition at MoMA, New York, in 1969, Richard Meier belonged to a group of architects known as the New York Five. Another member of the group, Peter Eisenman, has experimented with even more rigorous and complex design methods. In response to philosopher Jacques Derrida's claims that no meaning is ever stable or decidable and no system ever closed or pure, Eisenman developed a number of design methods – strictly speaking, he has produced a new method for each project – that would engage both formal issues and non-architectural information.

Scaling is a good example of Eisenman's techniques. The term scaling is taken from fractal geometry, which for Eisenman was analogous to Derrida's notion of deconstruction, according to which the dismantling of structures expands the limits of conceptual structures. In a fractal, the same or a similar figure is repeated at different scales, and no scale can be considered more real or fundamental than any other. This lack of an originary scale appealed to Eisenman because it evoked Derrida's claim that meaning has no originary source. Unlike some other architects, Eisenman never used fractals as such in his designs. Instead, he would choose a few drawings and overlay or "superpose" them in different

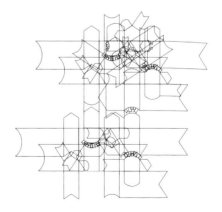

Fig. 83: Peter Eisenman, Biocenter Frankfurt/Main, diagram

Fig. 84: Peter Eisenman. The plan is generated from an overlay of the symbols for adenine, thymine, guanine and cytosine, the bases of DNA.

scales, as in his project for the Biocenter (1987) at the University in Frankfurt am Main. Out of the resulting, complex web of lines he then chose figures that combined fragments of different original images, but never drew any new lines. > Figs. 83 and 84

The first large deconstructivist building to be built, the Wexner Center for the Visual Arts in Columbus, Ohio (1983–89) by Eisenman and Richard Trott, questions conventional ideas of contextualism through

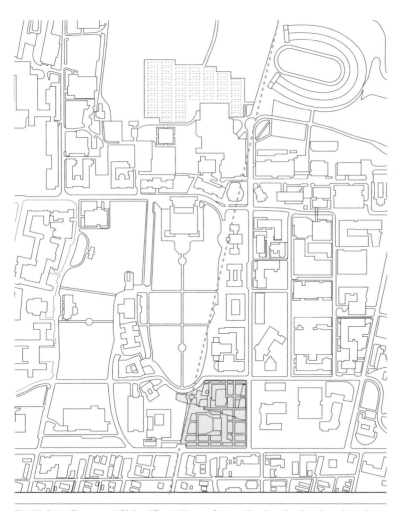

Fig. 85: Peter Eisenman & Richard Trott, Wexner Center site plan showing the axis to the football stadium

Fig. 86: Old Armory (left) and new Wexner Center on the Ohio State campus

the use of superposition and scaling. The building does not respond to its neighbors, the buildings of the Ohio State University campus, but instead responds to spatially or temporally remote physical conditions, somewhat like mosques directed towards Mecca. While the formal solution of the Wexner Center is based on the introduction of the street grid of the city into the university campus grid but 12.25° askew, the location and the main axes of the building are clearly determined by the football stadium several blocks away, and there is even a reference to a place 80 miles west of Columbus: at the northern end of the site, the complex but rational collision of the two grids is complicated through a replication of what Eisenman calls the Greenville Trace, a shear in the Jeffersonian grid, which was caused when the two groups of surveyors who were plotting the Ohio territory from opposite directions missed each other by a mile. Even when Eisenman imitates older buildings at the site, the referents are temporally distanced: the postmodern turrets recreate the forms of a 19th-century campus building, the "Old Armory", which had been torn down in 1958. These different materials are recorded in diagrammatic drawings duplicated in four different scales and overlaid to create a complex weave. Out of this composite drawing, Eisenman then chose a few lines to suggest fragments of the originals. > Figs. 85–87

MORPHING, FOLDING AND ANIMATE FORM

In addition to *superposition* and *scaling*, Eisenman also experimented with other ways of manipulating given images. In the late 1980s, an inexpensive software program briefly popularized the technique of *morphing*. Two or more images were chosen, essential points selected in each, and then one would be gradually transformed into the other; the design

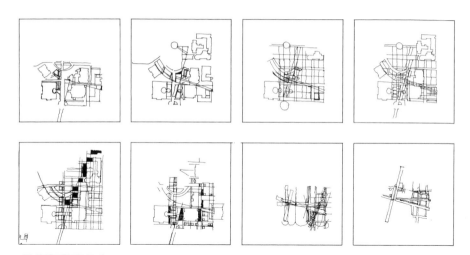

Fig. 87: Peter Eisenman, Wexner Center, design diagrams

would usually represent some point in the middle of the transformation process, where the qualities of the original images were no longer recognizable. In order to characterize the approach of their office, UN Studio, Ben van Berkel and Caroline Bos have often referred to a morphed image by artist Daniel Lee, the "manimal", which is a fusion of the images of a lion, a snake and a human face.

In a more serious sense, morphing was linked to the work of biologist D'Arcy Wentworth Thompson, who suggested in 1916 that the development of scientific morphology had been thwarted by a psychological tendency to see organic forms only in terms of their deviation from Euclidian geometry, rather than appreciating the similarities. Architect Greg Lynn's design for the Welsh National Opera House (1994), in Cardiff, Wales, is an example of a similar variation of core forms.

Another technique, known as *folding,* was sometimes understood literally as a kind of origami – e.g. Frankfurt Rebstock Park project (1991) by Eisenman – and sometimes as an adaptation of catastrophe or chaos theory. > Figs. 88 and 89

Originally, René Thom developed catastrophe theory as a way of describing biological morphogenesis mathematically. Especially in the 1990s, some followers of Eisenman tried to apply such theories to generating form. Lynn, for example, called for "animate form", which he

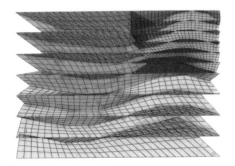

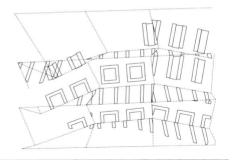

Fig. 88: Peter Eisenman, Rebstock Park, diagram showing the folding of the site

Fig. 89: Peter Eisenman, Rebstock Park, diagram showing the folding of volumes

developed by mapping external information, such as changing lighting conditions, in order to produce emergent, unforeseen, unpredictable, dynamic, and novel organizations in architecture.

One of the main ideas behind Greg Lynn's unrealized project (2001) for the Hydrogen House in Vienna was that it would register the movements of cars and of the sun in continuous topological surfaces composed of splines, smooth curves generated by a computer program. Much like the chronophotographs of Étienne-Jules Marey in the late 19th century, the Hydrogen House suggests movement by arranging variations of a form in a linear sequence. The form is not animate in the sense of the building actually moving, but its surface maps the movements from its environment. > Fig. 90

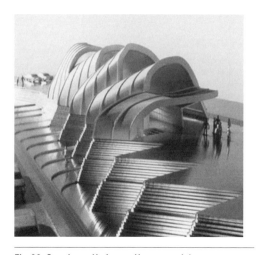

Fig. 90: Greg Lynn, Hydrogen House, model

DATASCAPE

The generative impulse in architectural method has been developed in yet another direction by Dutch architect Winy Maas, a partner in the firm MVRDV, with his concept of "datascapes", which combines deconstructive systems and design research approaches, often with a touch of irony.

The idea of datascapes is a delicate balance between unbridled technological reason and the ridicule or critique of modernism. Maas has taken as starting points, for instance, regulations in the Dutch building code, but has also turned fire escape routes, noise distribution charts, or garbage disposal systems into designs. Rules and constraints are set up and extrapolated with an iron logic *ad absurdum*. The intention is to present the rule in a pure and unexpected form that goes beyond artistic intuition and known geometries.

A good example of the approach is the project Monuments Act 2 by MVRDV from 1996. The issue investigated was the densification of the old center of Amsterdam so that new buildings would not be visible from the street. The inner court of a typical 18th-century block with a density of 0.8 is filled with a maximal volume cut in accordance to sight lines from the surrounding streets. With the resulting pointed volume in the middle, the density of the block rises tenfold to an amazing 7.8. Obviously, such a datascape is not a normal design, for the volume only satisfies one parameter (invisibility from the street) and deliberately ignores all other regulations in the code as well as any functional demands. > Fig. 91

Fig. 91: MVRDV, Monuments Act, Amsterdam

DIAGRAMS

Instead of deconstructive devices or datascapes, Ben van Berkel and Caroline Bos recommend the use of *diagrams,* that is, abstract means of thinking about organization, relationships, and possible worlds. Following Gilles Deleuze they explain that "the diagrammatic or abstract machine is not representational. It does not represent an existing object or situation, but it is instrumental in the production of new ones." They employ many different kinds of diagrams, including flow charts, music notations, schematic drawings of industrial buildings, electrical switch diagrams found in technical manuals, reproductions of paintings, and random images. Irrespective of their origins, the diagrams are read by van Berkel and Bos as infrastructural maps of movement.

The Möbius House (1997) in Het Gooi, the Netherlands, embodies certain qualities of diagrams that van Berkel and Bos had used as starting points, including a Paul Klee drawing. Thus, the lines of circulation in the building curve in and out just as the lines in Klee's diagram do. The name suggests a different diagram, the Möbius strip, which is a one-sided topological surface. A physical model of the Möbius strip is easy to make out of a long strip of paper by turning one end over and connecting the ends into a ring. Van Berkel und Bos do not, however, reproduce the topological qualities of the Möbius diagram, but interpret it as involving the mutation of dialectic opposites. In the Möbius House, the facades become inner walls, and glass and concrete swap places with every change in direction; programmatically, work connects to leisure; structurally, loadbearing elements transform into non-loadbearing ones.

The house was conceived as a 24-hour cycle of living, working, and sleeping with two intertwining paths that trace how two people can live together, yet apart, meeting at points of shared spaces. > Figs. 92 and 93

Architect and theorist Douglas Graf has developed a whole design theory based on the notion of non-representational diagrams. For Graf, the diagram is a device that negotiates between typologies that identify the components of an architectural composition, between the specific qualities of the particular building versus the general qualities that constitute a specifically architectural discourse, and between the stasis of configuration versus the dynamism of operation. To understand what Graf means it is best to discuss a specific case, for example his reading of the unbuilt Familian House project (1978) by Frank Gehry. Graf does not attempt to reconstruct Gehry's intentions, but to clarify how the plan entertains a game of center vs. edge, and openness vs. closedness. > Fig. 94

The house consists of two main elements, a square pavilion and a linear bar. The pavilion can be seen as a center framed by a perimeter, as defined by the larger volume. Let us first look at the bar more closely. > Fig. 95 While an infinite line can be thought of as made of identical points, a finite line is not as homogeneous: the end points of the line are different from the others, and the ends further imply a center (a).

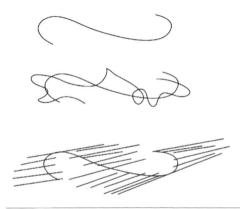

Fig. 92: UN Studio, drawing by Paul Klee as diagram for Möbius House

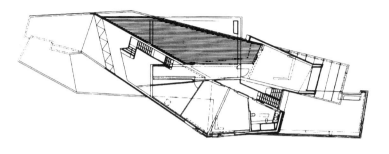

Fig. 93: UN Studio, Möbius House

Gehry's bar acknowledges these different conditions existing in any linear structure. He marks the center of the bar with a void and calls attention to the short ends by articulating them very differently: one is closed and perfect, the other, open if not disintegrating (b). This opposition of open vs. closed is repeated on the long sides of the bar as well: while the side towards the cube is smooth and closed, the back facade has several elements, such as a balcony and a stair, cantilevering off it. The cantilevered elements define a layer that corresponds to the layer inside the bar, as indicated by the internal corridor. Gehry suggests, then, a symmetrical organization about the central hall but goes on to destabilize this symmetry > Fig. 94. The terrace at the end of the bar reappears as the cubic volume (c) around which new white-clad material is collected to make a square pavilion (d).

Uncharacteristically for Gehry, the design for the Familian House exhibits a consistent geometrical layout. The apparently random angles of the bridges and the rotated square come from a regular pentagon. By drawing two identical pentagons, we can relate the pavilion to the bar and the central hall. The central axis of the pavilion intersects with the central axis of the bar, if we consider the latter's full extension. The hall corresponds to the balcony at the end of the bar. The width of the bar is also related to its length in the same proportion: the diagonal of one half bar is parallel to the side of the pavilion (e).

While the style or the formal language of the Familian House represents Gehry's idiosyncratic brand of deconstructivism, the themes that the design addresses are perennial in architecture; Graf demonstrates that similar questions and answers can be found, for example, in the Upper City of Pergamon and in Le Corbusier's Notre-Dame du Haut chapel

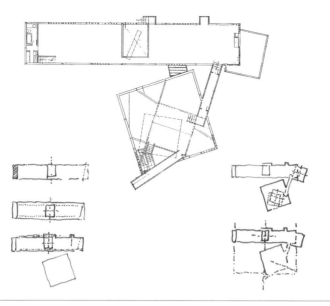

Fig. 94: Frank Gehry, Familian House, upper level plan

(1954) in Ronchamp. The same formal motifs, including those of emergent center, bi-nuclearity, edge vs. object, symmetry and its denial, are pursued in the design at many different scales.

PARAMETRIC DESIGN

In parametric design a set of independent parameters are chosen and systematically varied according to some criteria in order to arrive at not just one object but a series of variations. Usually, the parameters are given a geometrical interpretation.

Geometrically driven morphogenesis is usually discussed only in the context of recent computer-aided design systems. However, similar ideas were already explored at the very beginning of the period discussed in this book. Antoni Gaudí not only imagined strange organic shapes, but also used rational methods to develop and optimize them in the same way that current designers do. The best example of Gaudí's parametric design involves his famous hanging model, consisting of strings to which weights were connected, for the Colonia Güell chapel (1898–1915) near Barcelona. > Fig. 96

To understand how the method works, we need to know something about the catenary curve. If a perfectly flexible and absolutely homo-

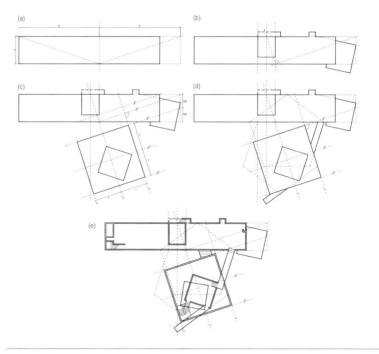

Fig. 95: Frank Gehry, Familian House, geometrical derivation of the plan

geneous string is suspended by its endpoints and not affected by any force other than gravity, it will assume the shape known as the catenary. Roughly, this shape is characteristic of suspension bridges, such as the Golden Gate in San Francisco, USA, where the weight of the roadway is divided equally along the overhead cables. In a chain hanging in a catenary curve, only tension forces exist. But if the shape is inverted by 180°, this creates an arch with only compression forces. In other words, the weight of the material acts along the curve without producing lateral forces. The Gateway Arch in St. Louis (1947–1966), USA, by Eero Saarinen, closely approximates the catenary curve. > Fig. 97

Although a catenary arch does not generate any lateral forces that could break it, it is only a two-dimensional structure, which could easily fall to one side if affected by the wind, for example. Hence, in architectural contexts the catenary arch is often extruded horizontally to make a vault, as in the great arch (400 AD) in Ctesiphon, Iran, > Fig. 98 or rotated along its central axis to make a catenoid dome, as inside the dome of St. Paul's Cathedral, London, designed by Christopher Wren in 1673.

Gaudí wanted to use the catenary curve to make an ideal, three-dimensional construction for a church. In ancient Roman architecture, arches, vaults, and domes were usually based on the circle. Gothic masons realized that by making the arch or the vault sharper, it was possible to extend the same span with less material, since the lateral forces were smaller. Still, the Gothic or pointed arch is not quite perfect, either: in order to counteract the lateral forces, Gothic masons added flying buttresses to the outside of the buildings. A catenary arch, a vault derived from the same shape, or a catenoid dome does not require any further support.

Gaudí's huge (4 × 6 m) hanging model was based on catenary curves. He started with an unweighted system of strings, and then began to vary their lengths, the points where they were connected, the weights that were hanging from them, etc. Every additional connection or weight would radically change the shape of the whole surface, exactly as in parametric design. Gaudí would have each configuration photographed and then make a final choice based on the spatial effects that he sought. This way, without a computer, he was able to determine with great accuracy extremely complex surfaces and be certain that the resulting geometry would act purely in compression when inverted. > Fig. 99

While Gaudí's hanging model is a way of finding the optimal construction for a given plan typology, the Foreign Office Architects (Farshid Moussavi and Alejandro Zaera-Polo) developed their design for the Yokohama Port Terminal (1996–2001) as the interaction of three factors: the program, the urban context, and the properties of the building materials.

Fig. 96: Antoni Gaudí, Colonia Güell, elevation of the crypt

Fig. 97: Eero Saarinen, Gateway Arch in St. Louis

Fig. 98: Arc at Ctesiphon, catenary vault

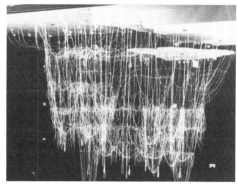

Fig. 99: Antoni Gaudí, hanging model with catenary curves

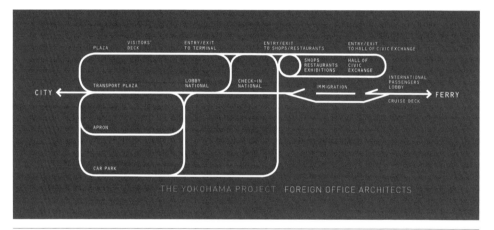

Fig. 100: Foreign Office Architects, Yokohama Port Terminal

Since the building is a ship terminal, the architects read the program largely in terms of circulation. Usually, a transportation building is a gateway for departures and arrivals, but the FOA wanted to define a field of movements without clear orientation. They separated the pedestrians, cars, trucks, and other kinds of circulation, and made each into a loop. > Fig. 100

Another of the architects' concerns was to create a hybrid between a shed and a pier, manipulating the ground plane so that it can also enclose spaces. Thus, the shapes are folded, as it were, from the

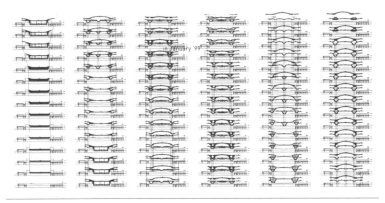

Fig. 101: Foreign Office Architects, Yokohama Port Terminal, sections

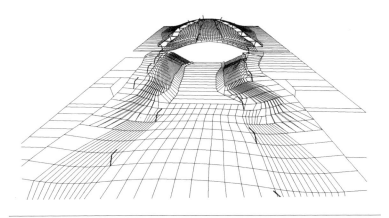

Fig. 102: Foreign Office Architects, Yokohama Port Terminal, ground plain

original pier to create a multilevel structure that seems to grow organically from the ground. The third task was to study the structural possibilities of the chosen materials, concrete, and glass. Of particular importance was the question of which spans and cantilevers were possible.
> Figs. 101 and 102

In parametric design, the architect does not impose a top-down architectural *gestalt* but works from bottom up with the cumulative calculation of the parameters, letting an unexpected form emerge from a computer algorithm. Such computerization of the design process has other implications as well. Instead of designing an individual object, the architect defines a basic building type, which can be adjusted to the particular context and modified according to the demands of a particular client. An example is the Variomatic House, proposed by Kas Oosterhuis in the late 1990s. The client was able to customize the design through a game-like interface, and the revised specifications could be sent directly to the companies producing the different components of the house. Parametric design can be combined with mass customization and computer-aided manufacturing, combining the benefits of cost-effective industrial production with the specific response to the client's needs, which had in fact been characteristic of traditional architectural design. By letting the client actively participate in the shaping of the architectural object, parametric methods open up a new chapter in the design of architecture.

In conclusion

Different design tasks call for different methods. When inserting a new block of housing into the old town fabric, it often makes sense to work contextually. By contrast, aleatory systems, including surrealist and deconstructivist methods, tend to produce unusual shapes that draw a lot of attention to themselves and are usually expensive to build. If a design that can be realized on a low budget is called for, a modular approach is a promising option. The modification of a typological or real precedent can be an effective way to make a confusing program understandable to the users. The users' involvement in the planning process can produce a good result if the users can be given the tools to choose from relevant alternative solutions. Parametric designs are most convincing when there is a reason to produce a series of similar elements with variations. Each method has its strengths and its limitations.

One of the methods is intuition. Some architects, such as Frank Lloyd Wright, boast of having received their designs in a dream, complete with all the details. Whether this is true or not, inspiration alone is not enough. Following the philosopher Karl Popper, we can distinguish between discovery and justification. Thus, Isaac Newton may have been inspired to formulate the law of gravity because an apple fell on his head, but what really matters is that his theory is justified because it agrees with the facts and other theories, and Newton was able to see that. As Louis Pasteur remarked in 1854, chance favors only the prepared mind.

What we call intuition is often better described as expertise: only someone who has internalized the knowledge of her field to the point that she can arrive at correct conclusions rapidly, without conscious deliberation. Intuition in this sense is indispensable to an architect, whether one uses a specific design method or not. The methods described above will offer you the means of developing a complex design without waiting desperately for inspiration, but they will not automatically tell you when you have reached a good solution. In order to recognize which design is the best, an architect needs to internalize the discourse of architecture and to understand the role of architecture in society. This understanding constitutes expertise in architecture.

Appendix

LITERATURE

Christopher Alexander: *A Pattern Language: Towns, Building, Construction,* Oxford University Press, New York 1977

Peter Eisenman: *Diagram Diaries,* Thames & Hudson, London 1999

Foreign Office Architects: *The Yokohama Project,* Actar, Barcelona 2002

Jacqueline Gargus: *Ideas of Order,* Kendall-Hunt, Dubuque, Iowa 1994

Douglas E. Graf: *Diagrams,* in: Perspecta Vol. 22, 1986, pp. 42–71.

Bill Hillier, Julienne Hanson: *The Social Logic of Space,* Cambridge University Press, Cambridge 1988

Greg Lynn: *Animate Form,* Princeton Architectural Press, New York 1999

William John Mitchell: *The Logic of Architecture: Design, Computation, and Cognition,* MIT Press, Cambridge, Mass. 1990

Elizabeth Martin: *Architecture as a Translation of Music,* Princeton Architectural Press, New York 1996

Mark Morris: *Automatic Architecture, Design from the Fourth Dimension,* University of North Carolina – College of Architecture, Charlotte 2006

Colin Rowe: *The Mathematics of the Ideal Villa and Other Essays,* MIT Press, Cambridge, Mass. 1988

Robert Venturi: *Mother's House. The Evolution of Vanna Venturi's House in Chestnut Hill,* Rizzoli, New York 1992

PICTURE CREDITS

Figures 35, 36, 61, 62, 63, 68, 85, 94, 95: Stefan Arbeithuber (Drawings)

Figures 20, 71, 72, 73, 74, 75, 78, 79, 97, 98: Kari Jormakka (Photographs, drawings)

Figures 19, 70, 80, 81: Claudia Kees (Drawings)

Figures 1, 5, 10, 14, 38, 82, 86: Dörte Kuhlmann (Photographs, Drawings)

Figures 70, 83, 84, 87, 88, 89, 91: Marta Neic (Redrawings)

Figures 11, 12, 17, 18, 21, 45, 51, 65, 66, 69, 101, 102: Alexander Semper (Redrawings)

Figures 13, 15, 16, 23, 26, 31, 32, 54, 55, 56, 58, 59, 60, 64, 70, 92, 93: Christina Simmel (Drawings, Redrawings)

Figures 2, 3, 4, 6, 7, 8, 9, 24, 25, 27, 28, 29, 30, 33, 34, 37, 39, 41, 42, 43, 44, 46, 47, 48, 49, 50, 52, 53, 57, 67, 76, 77, 90, 96, 99, 100: Bilderarchiv Institut für Architekturwissenschaften, TU Wien

THE AUTHORS
Kari Jormakka, 1959–2013, O. Univ. Prof. Dipl.-Ing. Dr. phil., professor
in the Department of Architectural Theory, University of
Technology Vienna
Oliver Schürer, author, curator, editor and senior scientist at
Department of Architectural Theory, University of Technology
Vienna
Dörte Kuhlmann, Ao. Univ. Prof. Dipl.-Ing. Dr.-Ing., Department
of Architectural Theory, University of Technology Vienna

Academic assistance
Gareth Griffiths, March., Liz. Tech., editor, Otaniemi University
of Technology, Finland
Alexander Semper, Dipl.-Ing. student assistant in the Department
of Architectural Theory, University of Technology Vienna

Drawings of diagrams
Claudia Kees
Stefan Arbeithuber

Editorial assistance
Christina Simmel, student assistant in the Department of Architectural
Theory, University of Technology Vienna
Marta Neic, student assistant in the Department of Architectural
Theory, University of Technology Vienna

Series editor: Bert Bielefeld
Concept: Bert Bielefeld, Annette Gref
English copy editing: Monica Buckland
Layout and cover design: Andreas Hidber
Typesetting and production: Amelie Solbrig

Library of Congress Cataloging-in-Publication data
A CIP catalog record for this book has been applied for at the Library of Congress.

Bibliographic information published by the German National Library
The German National Library lists this publication in the Deutsche Nationalbibliografie; detailed bibliographic data are available on the Internet at http://dnb.dnb.de.

This book is also available in a German (ISBN 978-3-7643-8462-3) and a French (ISBN 978-3-7643-8464-7) language edition.

© 2014 Birkhäuser Verlag GmbH, Basel
P.O. Box 44, 4009 Basel, Switzerland
Part of Walter de Gruyter GmbH, Berlin/Boston

Printed on acid-free paper produced from chlorine-free pulp. TCF ∞

Printed in Germany

ISBN 978-3-03821-520-2

9 8 7 6 5 4 3 2

www.birkhauser.com